Prabhakar Shukla et al.

Water quality assessment on river Ganga from Haridwar to Garhmukteshwar with a specific focus on point source discharges

GRIN Verlag

Bibliografische Information der Deutschen Nationalbibliothek:

Die Deutsche Bibliothek verzeichnet diese Publikation in der Deutschen National-
bibliografie; detaillierte bibliografische Daten sind im Internet über http://dnb.d-
nb.de/ abrufbar.

Imprint:

Copyright © 2014 GRIN Verlag GmbH
Druck und Bindung: Books on Demand GmbH, Norderstedt Germany
ISBN: 978-3-656-82823-5

This book at GRIN:

http://www.grin.com/en/e-book/282723/water-quality-assessment-on-river-ganga-
from-haridwar-to-garhmukteshwar

WATER QUALITY ASSESSMENT ON RIVER GANGA FROM HARIDWAR TO GARHMUKTESHWAR WITH A SPECIFIC FOCUS ON POINT SOURCE DISCHARGES

Mohit Chaudhary, Saurabh Mishra, Prabhakar Shukla, Arun kumar
Alternate Hydro Energy Center
Indian Institute of Technology
Roorkee, India

Abstract: There were many schemes for pollution abatement of River Ganga. Ganga Action Plan and thereafter NRCD and NGBRA were launched by Government of India. But these plans focused only on big cities and did not take into account pollution generated in the entire stretch/catchment. In their efforts small cities, semi urban settlements, Industries and pollution from rural sector were not addressed. As a result program had limited success and the improvement in water quality of River Ganga was limited. This study aims at studying the water quality assessment of point source discharges in river Ganga catchment from Haridwar to Garhmukteshwar (160 km) based on field visits , satellite data supported with details from survey of India topographical sheets, census data, district Industry data, water quality and discharge data. Catchment area of study stretch of river Ganga is about 5690 sq. km Water samples collected from drains and tributaries had been analyzed for pH, BOD COD, DO, TDS, Conductance, Nitrate, Heavy metals, Sulphate, Salinity, Turbidity, Faecal Coliform. These data are obtained by collecting water and wastewater samples during pre- and post-monsoon seasons during post monsoon water quality analysis (NOV 2013) and pre monsoon water quality analysis (Feb/March 2014) and are converted with National Sanitation Foundation Water Quality Index (NSFWQI) to achieve at a single value defining the water quality at selected locations. The result shows that the water quality of river is not good.

Keywords: Ganga action plan, river Ganga, water quality, NSFWQI.

INTRODUCTION

The Himalayas are the source of three major Indian rivers namely the Indus, the Ganga and the Brahmaputra. Ganga basin is the largest river basin in India in terms of catchment area, constituting 26% of the country's land mass (8,61,404 Sq. km) and supporting about 43% of its population (448.3 million as per 2001 census). The basin lies between East longitudes 73° 30' and 89° 0' and North latitudes of 22° 30' to 31° 30', covering an area of 1,086,000 sq. km, extending over India, Nepal and Bangladesh. About 79% area of Ganga basin is in India. River Ganga in India can be divided in to three sections: Upper reach from the origin to Narora, middle reach from Narora to Ballia and lower reach from Ballia to its delta. The Ganga River has been considered as the most sacred river of India in Puranas. It is called as Ganga Maa or Ganga ji (or reverend Ganga). The availabilities of abundant water resources, fertile soil, and suitable climate have given rise to a highly developed agriculture based civilization and one of the most densely populated regions of the world. The net sown area in the Ganga basin in India is around 44 million hectares (M-ha) and the net irrigated area is 23.41 M-ha. Rapidly increasing population, rising standards of living and exponential growth of industrialization and urbanization have exposed Ganga River in particular, to various forms of degradation. The deterioration in the water quality impacts the people immediately. Ganga, in some stretches, particularly during lean seasons has become unfit even for bathing. The threat of global climate change and the effect of glacial melt on Ganga flow, taking into account also the impacts of infrastructural projects in the upper reaches of the river, raise issues. In the Ganga basin, the flooding problem is mainly confined to the middle and terminal reaches. In 1985, the Government of India launched the Ganga Action Plan (GAP) GAP1and GAP 2) with the primary objective of cleaning the river. GAP has been a mixed success. Though good results were in many stretches but the problem of pollution was not fully addressed. A number of studies regarding pollution aspects of river Ganga and its tributaries have been carried out by different workers from AHEC, IIT Roorkee, Wildlife Institute of India (2012) Ministry of Environment and Forests (2009), Sravan Kumar Kota 2012, Prasanta Kar 2013 etc.

STUDY AREA

Ganga River from Bhimgauda barrage (Haridwar) to Garhmukteshwar has been considered for the study .This study catchment has catchment area of 5690 sq. km. The study stretch lies between East Longitude 78° 10' to 78°08 ' North Latitudes 29° 57' to 28° 46'. The length of stretch is 160 km. Barrage in study stretch are Bhimgauda barrage and Bijnor barrage. The Distance from Bhimgauda to Bijnor barrage is 82 km. The Population of stretch is 37.6 lakh (approx.). The figure 1 represents a line diagram of catchment under study and figure 2 represents the drainage map of sampling locations.

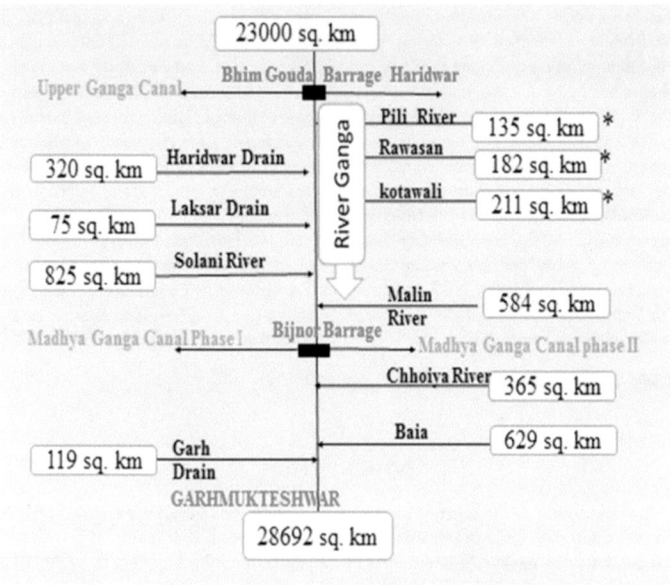

Fig: 1 Line diagram of study catchment (*Source*: www.nih.ernet.in)

Causes of Pollution in Stretch

Many rivers and drains joins the river stretch from Haridwar to Garhmukteshwar which are Solani river, Malin river, Pili river, Rawasan river, Kotawali river ,Chhoiya river , Baia ,Garh drain, Haridwar drain , Laksar drain. Some of these rivers bring in a lot of domestic/industrial pollution load. In addition, several towns, industries and agricultural activities contribute to the point and non-point pollution in this study stretch. Thus river flow and water quality are the key concerns in this stretch in addition to general degradation of river system and encroachment of river bed, sand mining, riverbed farming, open defecation etc. At many places dumping of solid waste and other materials used for religious purposes, washing of clothes, throwing partially burnt dead bodies add adverse effect on aesthetics, water quality and Aquatic Life.

2

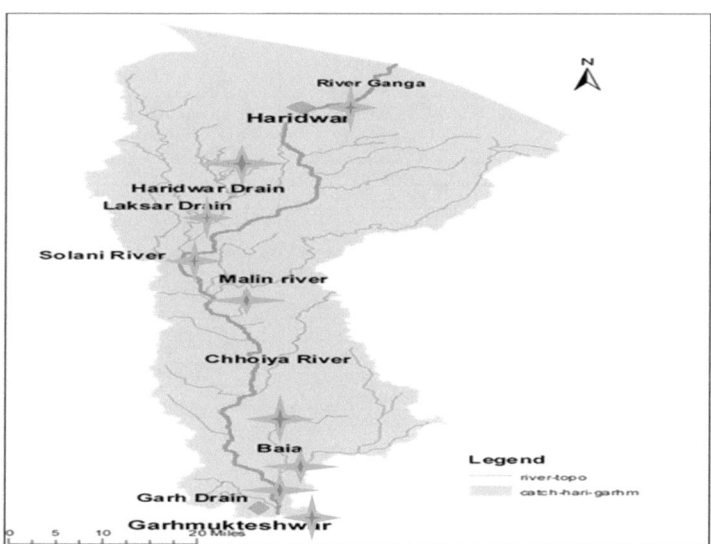

Fig: 2 Drainage map showing sampling locations in study catchment (*Source*: Arc_GIS image, Own Data)

METHODOLOGY

Samples were collected as per (APHA 2005) and test analysis was done in laboratory at AHEC, IITR. In brief each sample represent a composite sample, water samples collected were mixed from one-third, one-half and two-third width of a river along a transect. Samples were collected from 15-30 cm of water surface using standard water sampler (Hydro bias Germany), and kept in large airtight plastic ice-cold containers at 40^0C, then transported to laboratory within 8 hours of their collection. Samples were kept in different bottles with specific preservative for metal analysis and physiochemical analysis. D.O, pH and conductance were measured at the site with portable meter. The physiochemical analyses were done as per the standard methods (APHA 2005). For each period, three replicates of the water samples were collected and the average of three samples for each parameter to be studied was considered as one reading. The table 1 represents the analytical techniques used during testing.

Table: 1 Analytical Technique Used

Sn No.	Parameters	Analytical Techniques	Instrument used
1.	pH	Electrometric	Digital pH meter
2.	BOD	Volumetric	Dilution technique and keeping the sample at 20^0C for 5-days followed by winkers method
3.	COD	Electrometric	Digital COD meter
4.	Dissolved oxygen	Electrometric	Digital DO meter
5.	Total Dissolved solids	Gravimetric	Evaporation of filtrate (Whatman paper no. 44) at 103^0-105^0C
6.	Conductance	Electrometric	Digital conductivity meters
7	Hardness	Volumetric	Titrimetric method using EDTA soln. (0.01 M)
9	Nitrate	Colorimetric	Hach make Spectrophotometer
11.	Sulphate	Volumetric	Precipitate as BaSO4
12.	Salinity	Volumetric	Hach make Spectrophotometer
13.	Turbidity	Volumetric	Turbidometer
14.	Faecal Coliform	Volumetric	

(Source :MoEF , [13])

RESULT

Many rivers and drains joins the Ganga river stretch from Haridwar to Garhmukteshwar which are Bhimgauda U/S Haridwar drain, Laksar drain, Solani river, Malin river, Chhoiya River, Baia, Garh drain. Laboratory water quality analysis of tributaries of river Ganga and drains are shown in table 4 and 5 of post and pre monsoon sampling. NSWFQI online calculator has been used to calculate average water quality index using seven parameters DO, Feacal coliform, pH, BOD, Temperature change, Nitrate, and Turbidity. The weight factor for parameters used is represented in table 2. The water quality status is represented on 0-100 point scale are categorized on different range as excellent, good, medium, bad, and very bad quality as represented in table 3. In table 6 WQI calculated for post monsoon water quality analyses shows that except Bhimgauda U/S which has meddium water quality, rest other tributaries and drain has bad water quality status during discharge in Ganga River. In table 7 which represent the pre monsoon water quality index indicates that all drains and tributaries of river Ganga has bad water quality status during discharge.

Formula used in calculation of average WQI:

$$avg.wqi = Wi \times Qi$$

Where: avg.wqi - water quality index; Wi- weight of factor parameter; Xr- quality index of individual parameter.

$$Qi = Wi \times Vi$$

Vi –laboratory analyses value of individual parameter

Table: 2 Water Quality Factors and Weights (Ref [9])

Factors	Weight (Wi)
DO	0.17
Feacal coliform	0.16
pH	0.11
BOD	0.11
Temperature change	0.10
Nitrate	0.10
Turbidity	0.08

Table: 3 Water Quality Index Legends (Ref [9])

Range	Quality
90-100	Excellent
70-90	Good
50-70	Medium
25-50	Bad
0-25	Very bad

Table: 4 Post monsoon water quality analyses (NOV 2013)   (Source : Own Data)

Name of river /drain	DO (mg/l)	BOD (mg/l)	COD (mg/l)	Temp. (°C)	PH	Turbidity (NTU)	TDS (mg/l)	Salinity (parts per thousand) (‰)	Conductivity (µS/cm)	Faecal * Coliform MPN/10ml	Fe (mg/l)	Cu (mg/l)	Cr (mg/l)	Nitrate (mg/l)
Bhimgauda U/S	9.35	2.65	38	12	7.1	4.78	108.6	0.1	227	11	1.04	0.48	0.14	5.6
Garh D/S	8.44	5.18	112	16	7.7	22.7	111.7	0.1	233	21	0.42	0.54	0.02	4.2
Malin	8.5	2.8	19	24.5	8.3	4.94	243	0.2	501	15	0.74	-	0.14	8.4
Chhoiya	1.37	60	791	22.7	7.9	354	1574	1.6	3.07	290	2.81	1.91	0.55	15.8
Baia	7.93	4.70	32	20	8.9	12.5	245	0.2	507	43	0.5	-	0.04	3.6
Solani	7.23	6.88	76	21.3	8.6	12.8	196.9	0.2	408	20	0.46	1.38	0.1	4.9
Garh Drain	5.87	9.00	69	19	8	2	254	0.3	525	75	0.45	1.39	0.13	6.2
Haridwar Drain	7.66	13.00	43	18.8	8.1	19	206	0.2	427	170	UR	0.9	0.01	2.4
Laksar Drain	6.9	16.00	76	21	7.9	21	296	0.3	610	170	UR	0.6	UR	1.8

Table: 5 Pre monsoon water quality analyses (Feb/March 2014)

Name of river /drain	DO (mg/l)	BOD (mg/l)	COD (mg/l)	Tem. (°C)	PH	Turbidity (NTU)	TDS (mg/l)	Salinity (parts per thousand) (‰)	Conductivity (µS/cm)	Faecal Coliform MPN/10ml)	Fe (mg/l)	Cu (mg/l)	Cr (mg/l)	Nitrate (mg/l)
Bhimgauda U/S	9.11	2.00	27	14.4	7.2	5.4	123	0.2	280	1100	1.14	0.62	0.14	5.9
Garh D/S	7.86	6.05	39	17.7	7.9	107	156.8	0.2	325	2100	0.13	0.94	0.09	3.7
Malin	7.58	5.4	21	21.9	8.1	11.9	259	0.3	534	1850	UR	1.10	0.10	7.6
Chhoiya	1.25	87	361	25.8	8.1	154	1743	1.8	3390	1290	0.88	2.57	0.22	11.7
Bhaia	8.07	6	19	20.3	8.7	4..66	197	0.2	197	1430	0.23	1.87	0.09	2.9
Solani	7.00	8.4	36	22	8.5	12.8	230	0.2	490	1300	0.46	1.38	0.11	4.9
Garh Drain	5.43	18.64	92	18.7	7.9	143	368	0.4	754	1800	0.66	1.99	0.13	8.7
Haridwar Drain	6.2	19	61	19	8.2	51	288	0.3	630	2700	UR	0.9	0.02	2.4
Laksar Drain	3.81	41	120	23.3	7.8	27	376	0.4	780	940	0.21	0.6	UR	1.7

(Source : Own Data)

Table: 6 Post monsoon water quality index analyses (NOV 2013) using NSFWQI calculator

Name of river /drain	Faecal coliform (QI)	DO (QI)	pH (QI)	Turbidity (QI)	BOD (QI)	Nitrate (QI)	Temperature (QI)	Avg.WQI
Bhimgauda U/S	70	7	90	86	69	62	36	56
Garh D/S	63	6	91	59	55	69	29	50
Malin	67	6	73	86	68	55	16	50
Chhoiya	34	3	87	5	5	42	18	27
Bhaia	54	6	52	71	57	78	22	45
Solani	63	6	63	70	46	66	20	45
Garh Drain	47	5	84	93	38	60	24	45
Haridwar Drain	39	6	80	62	25	93	24	40
Laksar Drain	39	5	87	60	18	95	20	42

(Source : Own Data)

Table: 7 Pre monsoon water quality index analyses (Feb/March 2014) using NSFWQI calculator

Name of river /drain	Faecal coliform (QI)	DO (QI)	pH (QI)	Turbidity (QI)	BOD (QI)	Nitrate (QI)	Temperature (QI)	Avg.WQI
Bhimgauda U/S	22	7	92	85	80	61	32	48
Garh D/S	18	6	87	5	50	76	26	36
Malin	19	6	80	72	54	56	19	39
Chhoiya	21	3	80	5	5	48	15	25
Bhaia	20	6	59	87	51	91	21	41
Solani	21	6	66	70	40	66	19	36
Garh Drain	19	5	87	5	14	54	24	28
Haridwar Drain	17	5	77	38	13	93	24	34
Laksar Drain	23	4	90	55	5	95	17	37

(Source : Own Data)

CONCLUSION

As per physiochemical and heavy metal analyses of tributaries and drains of river Ganga during discharge, the water mixing into river is bad quality. Water quality analysis of drains falling into river stretch revealed that Chhoiya River was highly polluted river around, while water quality of other drains is variable depending on time of disposal of industrial waste into drains in different seasons. Impact assessment of point sources on river water quality using NSFWQI revealed that presently water in river is fulfills the requirement of outdoor bathing, but at various location in the river stretch water quality is likely to be affected due to severe degradation of water quality (BOD >3 and DO <6) and would not be suitable for bathing. It is recommended that waste water to be discharge in river should be treated, and proper minimal flow of river is to be maintained by supplying fresh water from other sources.

8

REFERENCES

1. Impact of hydro projects on water quality (Sravan Kumar Kota 2012), M.tech Dissertation AHEC IIT Roorkee.
2. Environmental Management Plan of River Brahmani (Prasanta Kar 2013) Mtech Dissertation AHEC IIT Roorkee.
3. Wildlife Institute of India (2012): Assessment of cumulative impacts of hydroelectric projects on terrestrial biodiversity in Alaknanda and Bhagirathi Basins, Uttarakhand.
4. Status Paper on river Ganga NRCD, Ministry of Environment and Forests, august 2009.
5. Assessment of Cumulative Impact of Hydropower Projects in Alaknanda and Bhagirathi Basins, by AHEC page no. 4-88. (2011)
6. Canter, L.W. and Clark, E.R "NEPA Effectiveness – A Survey of Academics", Environmental Impact Assessment Review, Vol. 17(5) page no. 313-327 (1997).
7. Culp JM, Cash KJ, Wrona FJ ; Integrated assessment of ecosystem integrity of large
8. Northern rivers: the northern river basins study example, J Aquat Ecosyst Stress Recovery page no. 8-15 (2000).
9. http://www.water-research.net/index.php/water-treatment/water-monitoring/monitoring-the-quality-of-surfacewaters: assessed on 25/09/2114.
10. http://www.nih.ernet.in/rbis/basin%20maps/ganga_about.htm (Hydrology and Water Resources Information System for India). Assessed on 25/01/2014.
11. http://ueppcb.uk.gov.in/pages/display/109-categorization,Official Website of Environment Protection and Pollution Control Board, Government of Uttarakhand.
12. http://upgov.nic.in/upstateglance.aspx, Official web site Government of Uttar Pradesh for Growth Rate During 2001 -2011.
13. www.moef.nic.in   (Ministry of Environment and Forests) assessed on 12/01/14
14. Detailed project report on sewerage and sewage treatment plant for Narora and Anupsahar towns (Dec.2013)
15. WWF REPORT, Assessment of environmental flows for the upper Ganga Basin; IWMI [2012].

9